MATH
WORKBOOK

ADDITION

1. 163
    + 145
    _____

2. 176
    + 23
    _____

3. 119
    + 97
    _____

4. 94
    + 58
    _____

5. 76
    + 195
    _____

6. 123
    + 155
    _____

7. 32
    + 85
    _____

8. 101
    + 94
    _____

9. 81
    + 15
    _____

10. 65
    + 135
    _____

11. 51
    + 196
    _____

12. 38
    + 74
    _____

13. 32
    + 141
    _____

14. 82
    + 188
    _____

15. 69
    + 60
    _____

16. 3
    + 71
    _____

17. 182
    + 32
    _____

18. 139
    + 116
    _____

19. 133
    + 115
    _____

20. 194
    + 21
    _____

21. 127
    + 199
    _____

22. 75
    + 177
    _____

23. 153
    + 134
    _____

24. 125
    + 64
    _____

1. **13**
    + 59

2. **6**
    + 59

3. **122**
    + 68

4. **115**
    + 106

5. **14**
    + 34

6. **68**
    + 12

7. **29**
    + 178

8. **73**
    + 57

9. **5**
    + 173

10. **7**
    + 33

11. **128**
    + 22

12. **3**
    + 30

13. **17**
    + 60

14. **182**
    + 99

15. **159**
    + 121

16. **117**
    + 135

17. **111**
    + 62

18. **14**
    + 143

19. **127**
    + 169

20. **47**
    + 52

21. **194**
    + 88

22. **168**
    + 186

23. **40**
    + 181

24. **194**
    + 25

1. $126 + 164$

2. $30 + 54$

3. $46 + 27$

4. $182 + 44$

5. $59 + 168$

6. $108 + 46$

7. $112 + 149$

8. $148 + 179$

9. $13 + 23$

10. $91 + 146$

11. $132 + 178$

12. $165 + 175$

13. $140 + 78$

14. $77 + 108$

15. $89 + 29$

16. $52 + 61$

17. $198 + 170$

18. $45 + 30$

19. $103 + 196$

20. $154 + 40$

21. $3 + 10$

22. $12 + 117$

23. $114 + 109$

24. $185 + 164$

1. **61**
    + 65

2. **141**
    + 135

3. **166**
    + 60

4. **61**
    + 43

5. **87**
    + 118

6. **116**
    + 175

7. **59**
    + 106

8. **12**
    + 27

9. **63**
    + 101

10. **25**
    + 54

11. **89**
    + 181

12. **91**
    + 128

13. **54**
    + 174

14. **198**
    + 66

15. **66**
    + 49

16. **61**
    + 68

17. **40**
    + 147

18. **91**
    + 185

19. **41**
    + 175

20. **182**
    + 148

21. **196**
    + 78

22. **88**
    + 37

23. **163**
    + 174

24. **180**
    + 24

1.   114
   + 85

2.   51
   + 26

3.   93
   + 69

4.   162
   + 86

5.   193
   + 184

6.   17
   + 51

7.   160
   + 62

8.   158
   + 20

9.   18
   + 20

10.   17
   + 33

11.   96
   + 189

12.   91
   + 196

13.   102
   + 50

14.   175
   + 112

15.   39
   + 73

16.   6
   + 26

17.   45
   + 193

18.   93
   + 33

19.   178
   + 116

20.   188
   + 111

21.   55
   + 199

22.   2
   + 78

23.   185
   + 131

24.   87
   + 199

1.   23
+ 114

2.   170
+ 103

3.   188
+   27

4.   11
+ 176

5.   156
+ 177

6.   11
+ 117

7.   170
+   15

8.   7
+ 97

9.   1
+ 169

10.   36
+ 127

11.   196
+ 141

12.   21
+ 200

13.   157
+   81

14.   94
+ 187

15.   129
+ 123

16.   37
+ 166

17.   93
+ 192

18.   178
+ 106

19.   183
+ 100

20.   138
+ 172

21.   46
+ 143

22.   95
+ 115

23.   45
+ 132

24.   157
+ 175

1.   **31**   2.   **79**   3.   **44**   4.   **155**
+ 70   **+ 64**   **+ 114**   **+ 36**

5. **119**   6. **194**   7. **153**   8. **66**
+ 124   **+ 30**   **+ 50**   **+ 185**

9. **83**   10. **13**   11. **155**   12. **47**
+ 99   **+ 22**   **+ 34**   **+ 30**

13. **94**   14. **171**   15. **109**   16. **146**
+ 65   **+ 199**   **+ 180**   **+ 94**

17. **199**   18. **165**   19. **32**   20. **41**
+ 69   **+ 43**   **+ 60**   **+ 182**

21. **16**   22. **194**   23. **31**   24. **85**
+ 117   **+ 14**   **+ 34**   **+ 103**

1. $53 + 39$

2. $198 + 49$

3. $32 + 50$

4. $191 + 75$

5. $173 + 80$

6. $20 + 79$

7. $84 + 21$

8. $195 + 108$

9. $112 + 163$

10. $19 + 175$

11. $90 + 62$

12. $44 + 66$

13. $194 + 62$

14. $30 + 128$

15. $3 + 195$

16. $22 + 153$

17. $84 + 72$

18. $82 + 69$

19. $98 + 113$

20. $174 + 176$

21. $67 + 100$

22. $14 + 72$

23. $137 + 54$

24. $56 + 165$

1. 22 oranges are in the basket. 33 more oranges are put in the basket. How many oranges are in the basket now?

....

2. David has 17 marbles and Steven has 19 marbles. How many marbles do David and Steven have together?

....

3. Janet has 38 more apples than Jackie. Jackie has 40 apples. How many apples does Janet have?

....

4. 22 plums were in the basket. 19 are red and the rest are green. How many plums are green?

....

5. Some balls were in the basket. 12 more balls were added to the basket. Now there are 21 balls. How many balls were in the basket before more balls were added?

....

6. 3 red peaches and 36 green peaches are in the basket. How many peaches are in the basket?

....

7. 13 pears were in the basket. More pears were added to the basket. Now there are 20 pears. How many pears were added to the basket?

....

1. 24 marbles are in the basket. 12 more marbles are put in the basket. How many marbles are in the basket now?

2. Sandra has 22 more peaches than Janet. Janet has 17 peaches. How many peaches does Sandra have?

3. 13 red oranges and 2 green oranges are in the basket. How many oranges are in the basket?

4. Adam has 27 apples and Billy has 9 apples. How many apples do Adam and Billy have together?

5. 60 plums were in the basket. 39 are red and the rest are green. How many plums are green?

6. 11 balls were in the basket. More balls were added to the basket. Now there are 25 balls. How many balls were added to the basket?

7. Some pears were in the basket. 29 more pears were added to the basket. Now there are 56 pears. How many pears were in the basket before more pears were added?

1. 25 oranges are in the basket. 7 more oranges are put in the basket. How many oranges are in the basket now?

2. Steven has 11 marbles and Adam has 25 marbles. How many marbles do Steven and Adam have together?

3. 39 pears were in the basket. 4 are red and the rest are green. How many pears are green?

4. 26 plums were in the basket. More plums were added to the basket. Now there are 50 plums. How many plums were added to the basket?

5. Some balls were in the basket. 10 more balls were added to the basket. Now there are 45 balls. How many balls were in the basket before more balls were added?

6. Jackie has 19 more apples than Audrey. Audrey has 31 apples. How many apples does Jackie have?

7. 37 red peaches and 4 green peaches are in the basket. How many peaches are in the basket?

1. Some plums were in the basket. 7 more plums were added to the basket. Now there are 33 plums. How many plums were in the basket before more plums were added?

   ....

2. 33 red peaches and 8 green peaches are in the basket. How many peaches are in the basket?

   ....

3. Jackie has 20 more apples than Audrey. Audrey has 22 apples. How many apples does Jackie have?

   ....

4. Adam has 12 oranges and David has 27 oranges. How many oranges do Adam and David have together?

   ....

5. 23 balls were in the basket. 12 are red and the rest are green. How many balls are green?

   ....

6. 17 pears were in the basket. More pears were added to the basket. Now there are 37 pears. How many pears were added to the basket?

   ....

7. 23 marbles are in the basket. 3 more marbles are put in the basket. How many marbles are in the basket now?

   ....

SUBTRACTION

1. 141
    - 99

2. 68
    - 12

3. 75
    - 49

4. 77
    - 15

5. 148
    - 12

6. 75
    - 45

7. 158
    - 119

8. 182
    - 112

9. 132
    - 11

10. 11
    - 11

11. 71
    - 21

12. 125
    - 75

13. 133
    - 102

14. 44
    - 29

15. 102
    - 39

16. 66
    - 62

17. 134
    - 74

18. 137
    - 47

19. 96
    - 90

20. 116
    - 58

21. 94
    - 43

22. 71
    - 40

23. 179
    - 30

24. 34
    - 26

1.   **130**
  - 76

2.   **153**
  - 122

3.   **49**
  - 33

4.   **16**
  - 12

5.   **130**
  - 48

6.   **197**
  - 144

7.   **110**
  - 40

8.   **117**
  - 49

9.   **159**
  - 83

10.   **165**
  - 118

11.   **188**
  - 145

12.   **168**
  - 55

13.   **93**
  - 30

14.   **86**
  - 44

15.   **59**
  - 36

16.   **188**
  - 92

17.   **170**
  - 149

18.   **86**
  - 30

19.   **68**
  - 51

20.   **16**
  - 14

21.   **125**
  - 39

22.   **41**
  - 21

23.   **72**
  - 16

24.   **162**
  - 129

1. $\begin{array}{r} 103 \\ - 23 \\ \hline \end{array}$
2. $\begin{array}{r} 41 \\ - 31 \\ \hline \end{array}$
3. $\begin{array}{r} 163 \\ - 76 \\ \hline \end{array}$
4. $\begin{array}{r} 199 \\ - 57 \\ \hline \end{array}$

5. $\begin{array}{r} 93 \\ - 56 \\ \hline \end{array}$
6. $\begin{array}{r} 155 \\ - 42 \\ \hline \end{array}$
7. $\begin{array}{r} 93 \\ - 53 \\ \hline \end{array}$
8. $\begin{array}{r} 46 \\ - 16 \\ \hline \end{array}$

9. $\begin{array}{r} 98 \\ - 75 \\ \hline \end{array}$
10. $\begin{array}{r} 113 \\ - 100 \\ \hline \end{array}$
11. $\begin{array}{r} 76 \\ - 14 \\ \hline \end{array}$
12. $\begin{array}{r} 103 \\ - 12 \\ \hline \end{array}$

13. $\begin{array}{r} 48 \\ - 13 \\ \hline \end{array}$
14. $\begin{array}{r} 196 \\ - 140 \\ \hline \end{array}$
15. $\begin{array}{r} 198 \\ - 48 \\ \hline \end{array}$
16. $\begin{array}{r} 191 \\ - 64 \\ \hline \end{array}$

17. $\begin{array}{r} 194 \\ - 14 \\ \hline \end{array}$
18. $\begin{array}{r} 185 \\ - 45 \\ \hline \end{array}$
19. $\begin{array}{r} 106 \\ - 91 \\ \hline \end{array}$
20. $\begin{array}{r} 173 \\ - 45 \\ \hline \end{array}$

21. $\begin{array}{r} 93 \\ - 12 \\ \hline \end{array}$
22. $\begin{array}{r} 112 \\ - 75 \\ \hline \end{array}$
23. $\begin{array}{r} 166 \\ - 125 \\ \hline \end{array}$
24. $\begin{array}{r} 135 \\ - 118 \\ \hline \end{array}$

1. 177
    - 51
    ....

2. 93
    - 36
    ....

3. 110
    - 96
    ....

4. 47
    - 24
    ....

5. 87
    - 19
    ....

6. 146
    - 16
    ....

7. 49
    - 41
    ....

8. 129
    - 102
    ....

9. 68
    - 38
    ....

10. 124
    - 52
    ....

11. 77
    - 18
    ....

12. 154
    - 62
    ....

13. 140
    - 59
    ....

14. 17
    - 13
    ....

15. 117
    - 38
    ....

16. 59
    - 23
    ....

17. 63
    - 33
    ....

18. 64
    - 35
    ....

19. 154
    - 67
    ....

20. 182
    - 122
    ....

21. 190
    - 25
    ....

22. 74
    - 34
    ....

23. 76
    - 30
    ....

24. 192
    - 123
    ....

1. $\begin{array}{r} 121 \\ -\ 99 \\ \hline \end{array}$
2. $\begin{array}{r} 89 \\ -\ 29 \\ \hline \end{array}$
3. $\begin{array}{r} 158 \\ -\ 63 \\ \hline \end{array}$
4. $\begin{array}{r} 14 \\ -\ 12 \\ \hline \end{array}$

5. $\begin{array}{r} 89 \\ -\ 28 \\ \hline \end{array}$
6. $\begin{array}{r} 141 \\ -\ 20 \\ \hline \end{array}$
7. $\begin{array}{r} 41 \\ -\ 15 \\ \hline \end{array}$
8. $\begin{array}{r} 188 \\ -\ 147 \\ \hline \end{array}$

9. $\begin{array}{r} 43 \\ -\ 27 \\ \hline \end{array}$
10. $\begin{array}{r} 185 \\ -\ 107 \\ \hline \end{array}$
11. $\begin{array}{r} 156 \\ -\ 116 \\ \hline \end{array}$
12. $\begin{array}{r} 196 \\ -\ 37 \\ \hline \end{array}$

13. $\begin{array}{r} 66 \\ -\ 21 \\ \hline \end{array}$
14. $\begin{array}{r} 109 \\ -\ 100 \\ \hline \end{array}$
15. $\begin{array}{r} 110 \\ -\ 29 \\ \hline \end{array}$
16. $\begin{array}{r} 68 \\ -\ 44 \\ \hline \end{array}$

17. $\begin{array}{r} 99 \\ -\ 27 \\ \hline \end{array}$
18. $\begin{array}{r} 110 \\ -\ 63 \\ \hline \end{array}$
19. $\begin{array}{r} 178 \\ -\ 14 \\ \hline \end{array}$
20. $\begin{array}{r} 196 \\ -\ 33 \\ \hline \end{array}$

21. $\begin{array}{r} 18 \\ -\ 12 \\ \hline \end{array}$
22. $\begin{array}{r} 149 \\ -\ 92 \\ \hline \end{array}$
23. $\begin{array}{r} 21 \\ -\ 18 \\ \hline \end{array}$
24. $\begin{array}{r} 197 \\ -\ 109 \\ \hline \end{array}$

1.　113
　- 88

2.　134
　- 75

3.　14
　- 12

4.　35
　- 24

5.　48
　- 41

6.　86
　- 66

7.　124
　- 47

8.　75
　- 19

9.　21
　- 18

10.　145
　- 99

11.　25
　- 16

12.　68
　- 63

13.　84
　- 48

14.　136
　- 130

15.　35
　- 10

16.　27
　- 12

17.　141
　- 69

18.　93
　- 50

19.　137
　- 88

20.　163
　- 71

21.　27
　- 17

22.　84
　- 13

23.　146
　- 26

24.　129
　- 82

1. 180
    - 49

2. 119
    - 84

3. 200
    - 47

4. 49
    - 20

5. 178
    - 19

6. 46
    - 34

7. 30
    - 13

8. 86
    - 29

9. 83
    - 31

10. 16
    - 16

11. 101
    - 37

12. 144
    - 132

13. 194
    - 39

14. 70
    - 35

15. 24
    - 17

16. 146
    - 133

17. 198
    - 75

18. 15
    - 12

19. 175
    - 64

20. 39
    - 15

21. 110
    - 97

22. 129
    - 89

23. 198
    - 117

24. 25
    - 18

1.   35
  - 31

2.   88
  - 58

3.   137
  - 15

4.   184
  - 39

5.   34
  - 28

6.   107
  - 37

7.   144
  - 25

8.   25
  - 17

9.   104
  - 50

10.   171
  - 38

11.   138
  - 57

12.   55
  - 43

13.   137
  - 37

14.   71
  - 53

15.   27
  - 17

16.   88
  - 82

17.   87
  - 69

18.   99
  - 98

19.   35
  - 19

20.   187
  - 85

21.   152
  - 81

22.   136
  - 34

23.   63
  - 59

24.   144
  - 41

1. 25 plums were in the basket. Some of the plums were removed from the basket. Now there are 14 plums. How many plums were removed from the basket?

....

2. Some marbles were in the basket. 2 marbles were taken from the basket. Now there are 7 marbles. How many marbles were in the basket before some of the marbles were taken?

....

3. 8 peaches are in the basket. 3 peaches are taken out of the basket. How many peaches are in the basket now?

....

4. Jackie has 38 fewer oranges than Ellen. Ellen has 40 oranges. How many oranges does Jackie have?

....

5. 31 pears are in the basket. 2 are red and the rest are green. How many pears are green?

....

6. David has 4 balls. Billy has 12 balls. How many more balls does Billy have than David?

....

7. 36 apples were in the basket. Some of the apples were removed from the basket. Now there are 27 apples. How many apples were removed from the basket?

....

1. 25 marbles were in the basket. Some of the marbles were removed from the basket. Now there are 18 marbles. How many marbles were removed from the basket?

....

2. Some balls were in the basket. 19 balls were taken from the basket. Now there are 8 balls. How many balls were in the basket before some of the balls were taken?

....

3. 24 pears are in the basket. 10 are red and the rest are green. How many pears are green?

....

4. Sandra has 8 fewer apples than Marin. Marin has 17 apples. How many apples does Sandra have?

....

5. 6 peaches are in the basket. 4 peaches are taken out of the basket. How many peaches are in the basket now?

....

6. Paul has 12 plums. Steven has 36 plums. How many more plums does Steven have than Paul?

....

7. Some oranges were in the basket. 4 oranges were taken from the basket. Now there are 35 oranges. How many oranges were in the basket before some of the oranges were taken?

....

1. Donald has 16 balls. Billy has 37 balls. How many more balls does Billy have than Donald?

....

2. 20 plums are in the basket. 7 plums are taken out of the basket. How many plums are in the basket now?

....

3. Michele has 25 fewer oranges than Jennifer. Jennifer has 31 oranges. How many oranges does Michele have?

....

4. Some peaches were in the basket. 6 peaches were taken from the basket. Now there is 1 peach. How many peaches were in the basket before some of the peaches were taken?

....

5. 26 marbles were in the basket. Some of the marbles were removed from the basket. Now there are 7 marbles. How many marbles were removed from the basket?

....

6. 12 apples are in the basket. 3 are red and the rest are green. How many apples are green?

....

7. 4 pears are in the basket. 4 pears are taken out of the basket. How many pears are in the basket now?

....

1.   7 apples are in the basket. 4 apples are taken out of the basket. How many apples are in the basket now?

     ....

2.   Sandra has 29 fewer pears than Marcie. Marcie has 35 pears. How many pears does Sandra have?

     ....

3.   Donald has 8 plums. Jake has 33 plums. How many more plums does Jake have than Donald?

     ....

4.   Some balls were in the basket. 13 balls were taken from the basket. Now there are 12 balls. How many balls were in the basket before some of the balls were taken?

     ....

5.   21 marbles are in the basket. 18 are red and the rest are green. How many marbles are green?

     ....

6.   16 oranges were in the basket. Some of the oranges were removed from the basket. Now there are 5 oranges. How many oranges were removed from the basket?

     ....

7.   Michele has 22 fewer peaches than Marin. Marin has 40 peaches. How many peaches does Michele have?

     ....

MULTIPLICATION

1. **8** **× 4**	2. **2** **× 7**	3. **6** **× 4**	4. **6** **× 1**

1. **8**
    × 4

2. **2**
    × 7

3. **6**
    × 4

4. **6**
    × 1

5. **5**
    × 4

6. **3**
    × 4

7. **4**
    × 7

8. **2**
    × 8

9. **8**
    × 5

10. **7**
    × 4

11. **9**
    × 9

12. **4**
    × 2

13. **5**
    × 7

14. **4**
    × 5

15. **2**
    × 6

16. **4**
    × 8

17. **5**
    × 2

18. **5**
    × 5

19. **8**
    × 9

20. **6**
    × 3

21. **5**
    × 9

22. **4**
    × 4

23. **8**
    × 2

24. **1**
    × 2

1. $\begin{array}{r} 7 \\ \times\ 6 \\ \hline \end{array}$

2. $\begin{array}{r} 5 \\ \times\ 9 \\ \hline \end{array}$

3. $\begin{array}{r} 6 \\ \times\ 8 \\ \hline \end{array}$

4. $\begin{array}{r} 1 \\ \times\ 2 \\ \hline \end{array}$

5. $\begin{array}{r} 8 \\ \times\ 9 \\ \hline \end{array}$

6. $\begin{array}{r} 4 \\ \times\ 7 \\ \hline \end{array}$

7. $\begin{array}{r} 4 \\ \times\ 2 \\ \hline \end{array}$

8. $\begin{array}{r} 6 \\ \times\ 2 \\ \hline \end{array}$

9. $\begin{array}{r} 9 \\ \times\ 8 \\ \hline \end{array}$

10. $\begin{array}{r} 2 \\ \times\ 8 \\ \hline \end{array}$

11. $\begin{array}{r} 9 \\ \times\ 4 \\ \hline \end{array}$

12. $\begin{array}{r} 3 \\ \times\ 1 \\ \hline \end{array}$

13. $\begin{array}{r} 2 \\ \times\ 1 \\ \hline \end{array}$

14. $\begin{array}{r} 2 \\ \times\ 3 \\ \hline \end{array}$

15. $\begin{array}{r} 8 \\ \times\ 1 \\ \hline \end{array}$

16. $\begin{array}{r} 7 \\ \times\ 3 \\ \hline \end{array}$

17. $\begin{array}{r} 2 \\ \times\ 4 \\ \hline \end{array}$

18. $\begin{array}{r} 6 \\ \times\ 3 \\ \hline \end{array}$

19. $\begin{array}{r} 7 \\ \times\ 4 \\ \hline \end{array}$

20. $\begin{array}{r} 3 \\ \times\ 7 \\ \hline \end{array}$

21. $\begin{array}{r} 8 \\ \times\ 5 \\ \hline \end{array}$

22. $\begin{array}{r} 4 \\ \times\ 8 \\ \hline \end{array}$

23. $\begin{array}{r} 1 \\ \times\ 5 \\ \hline \end{array}$

24. $\begin{array}{r} 1 \\ \times\ 6 \\ \hline \end{array}$

1. $\begin{array}{r} 2 \\ \times\ 7 \\ \hline \end{array}$	2. $\begin{array}{r} 6 \\ \times\ 3 \\ \hline \end{array}$	3. $\begin{array}{r} 5 \\ \times\ 6 \\ \hline \end{array}$	4. $\begin{array}{r} 1 \\ \times\ 3 \\ \hline \end{array}$
5. $\begin{array}{r} 3 \\ \times\ 5 \\ \hline \end{array}$	6. $\begin{array}{r} 9 \\ \times\ 3 \\ \hline \end{array}$	7. $\begin{array}{r} 9 \\ \times\ 8 \\ \hline \end{array}$	8. $\begin{array}{r} 5 \\ \times\ 9 \\ \hline \end{array}$
9. $\begin{array}{r} 2 \\ \times\ 5 \\ \hline \end{array}$	10. $\begin{array}{r} 5 \\ \times\ 7 \\ \hline \end{array}$	11. $\begin{array}{r} 8 \\ \times\ 4 \\ \hline \end{array}$	12. $\begin{array}{r} 4 \\ \times\ 5 \\ \hline \end{array}$
13. $\begin{array}{r} 4 \\ \times\ 1 \\ \hline \end{array}$	14. $\begin{array}{r} 8 \\ \times\ 3 \\ \hline \end{array}$	15. $\begin{array}{r} 6 \\ \times\ 7 \\ \hline \end{array}$	16. $\begin{array}{r} 4 \\ \times\ 8 \\ \hline \end{array}$
17. $\begin{array}{r} 3 \\ \times\ 1 \\ \hline \end{array}$	18. $\begin{array}{r} 1 \\ \times\ 5 \\ \hline \end{array}$	19. $\begin{array}{r} 7 \\ \times\ 9 \\ \hline \end{array}$	20. $\begin{array}{r} 8 \\ \times\ 2 \\ \hline \end{array}$
21. $\begin{array}{r} 2 \\ \times\ 3 \\ \hline \end{array}$	22. $\begin{array}{r} 4 \\ \times\ 9 \\ \hline \end{array}$	23. $\begin{array}{r} 7 \\ \times\ 2 \\ \hline \end{array}$	24. $\begin{array}{r} 3 \\ \times\ 2 \\ \hline \end{array}$

1.　　**8**
　　× **6**

2.　　**3**
　　× **2**

3.　　**2**
　　× **4**

4.　　**3**
　　× **6**

5.　　**9**
　　× **5**

6.　　**3**
　　× **7**

7.　　**1**
　　× **8**

8.　　**2**
　　× **8**

9.　　**7**
　　× **2**

10.　　**5**
　　× **7**

11.　　**5**
　　× **4**

12.　　**4**
　　× **6**

13.　　**6**
　　× **3**

14.　　**4**
　　× **7**

15.　　**8**
　　× **4**

16.　　**5**
　　× **3**

17.　　**9**
　　× **7**

18.　　**9**
　　× **8**

19.　　**7**
　　× **7**

20.　　**2**
　　× **5**

21.　　**2**
　　× **2**

22.　　**1**
　　× **7**

23.　　**1**
　　× **6**

24.　　**5**
　　× **8**

1. 7
   × 4
   ....

2. 3
   × 5
   ....

3. 6
   × 1
   ....

4. 9
   × 7
   ....

5. 4
   × 9
   ....

6. 4
   × 3
   ....

7. 8
   × 5
   ....

8. 2
   × 2
   ....

9. 7
   × 1
   ....

10. 7
    × 5
    ....

11. 4
    × 6
    ....

12. 3
    × 4
    ....

13. 5
    × 2
    ....

14. 4
    × 2
    ....

15. 8
    × 3
    ....

16. 5
    × 5
    ....

17. 2
    × 9
    ....

18. 7
    × 3
    ....

19. 2
    × 8
    ....

20. 3
    × 1
    ....

21. 9
    × 4
    ....

22. 8
    × 1
    ....

23. 5
    × 9
    ....

24. 9
    × 2
    ....

1. $\begin{array}{r} 2 \\ \times\ 4 \\ \hline \end{array}$ 2. $\begin{array}{r} 2 \\ \times\ 5 \\ \hline \end{array}$ 3. $\begin{array}{r} 5 \\ \times\ 4 \\ \hline \end{array}$ 4. $\begin{array}{r} 5 \\ \times\ 8 \\ \hline \end{array}$

5. $\begin{array}{r} 3 \\ \times\ 4 \\ \hline \end{array}$ 6. $\begin{array}{r} 2 \\ \times\ 6 \\ \hline \end{array}$ 7. $\begin{array}{r} 3 \\ \times\ 5 \\ \hline \end{array}$ 8. $\begin{array}{r} 4 \\ \times\ 6 \\ \hline \end{array}$

9. $\begin{array}{r} 7 \\ \times\ 2 \\ \hline \end{array}$ 10. $\begin{array}{r} 8 \\ \times\ 8 \\ \hline \end{array}$ 11. $\begin{array}{r} 8 \\ \times\ 2 \\ \hline \end{array}$ 12. $\begin{array}{r} 5 \\ \times\ 2 \\ \hline \end{array}$

13. $\begin{array}{r} 4 \\ \times\ 3 \\ \hline \end{array}$ 14. $\begin{array}{r} 1 \\ \times\ 2 \\ \hline \end{array}$ 15. $\begin{array}{r} 4 \\ \times\ 4 \\ \hline \end{array}$ 16. $\begin{array}{r} 5 \\ \times\ 7 \\ \hline \end{array}$

17. $\begin{array}{r} 8 \\ \times\ 7 \\ \hline \end{array}$ 18. $\begin{array}{r} 4 \\ \times\ 7 \\ \hline \end{array}$ 19. $\begin{array}{r} 2 \\ \times\ 2 \\ \hline \end{array}$ 20. $\begin{array}{r} 8 \\ \times\ 4 \\ \hline \end{array}$

21. $\begin{array}{r} 3 \\ \times\ 2 \\ \hline \end{array}$ 22. $\begin{array}{r} 8 \\ \times\ 3 \\ \hline \end{array}$ 23. $\begin{array}{r} 3 \\ \times\ 7 \\ \hline \end{array}$ 24. $\begin{array}{r} 5 \\ \times\ 9 \\ \hline \end{array}$

1. 6
    × 3

2. 4
    × 7

3. 1
    × 7

4. 2
    × 7

5. 2
    × 8

6. 6
    × 8

7. 7
    × 7

8. 2
    × 5

9. 3
    × 8

10. 4
     × 5

11. 9
     × 9

12. 4
     × 3

13. 8
     × 2

14. 7
     × 4

15. 7
     × 2

16. 9
     × 6

17. 8
     × 7

18. 2
     × 3

19. 6
     × 4

20. 1
     × 5

21. 3
     × 3

22. 4
     × 4

23. 6
     × 1

24. 4
     × 6

1.   8
 × 5

2.   7
 × 8

3.   5
 × 4

4.   5
 × 9

5.   2
 × 3

6.   8
 × 7

7.   3
 × 2

8.   2
 × 8

9.   7
 × 5

10.   6
 × 2

11.   3
 × 3

12.   2
 × 2

13.   6
 × 4

14.   1
 × 3

15.   9
 × 4

16.   1
 × 5

17.   7
 × 7

18.   1
 × 8

19.   9
 × 7

20.   5
 × 7

21.   1
 × 1

22.   3
 × 4

23.   8
 × 6

24.   7
 × 4

1. Steven swims 6 laps every day. How many laps will Steven swim in 2 days?

....

2. Jake can cycle 10 miles per hour. How far can Jake cycle in 5 hours?

....

3. If there are 6 marbles in each box and there are 8 boxes, how many marbles are there in total?

....

4. Ellen has 5 times more peaches than Donald. Donald has 9 peaches. How many peaches does Ellen have?

....

5. Marcie's garden has 4 rows of pumpkins. Each row has 8 pumpkins. How many pumpkins does Marcie have in all?

....

6. If there are 9 apples in each box and there are 8 boxes, how many apples are there in total?

....

7. Marcie's garden has 10 rows of pumpkins. Each row has 7 pumpkins. How many pumpkins does Marcie have in all?

....

1. Sandra's garden has 2 rows of pumpkins. Each row has 5 pumpkins. How many pumpkins does Sandra have in all?

2. Sharon has 3 times more pears than Paul. Paul has 4 pears. How many pears does Sharon have?

3. If there are 2 oranges in each box and there are 6 boxes, how many oranges are there in total?

4. Marin swims 7 laps every day. How many laps will Marin swim in 10 days?

5. Paul can cycle 5 miles per hour. How far can Paul cycle in 7 hours?

6. Amy swims 10 laps every day. How many laps will Amy swim in 5 days?

7. Ellen's garden has 5 rows of pumpkins. Each row has 7 pumpkins. How many pumpkins does Ellen have in all?

1. Marin has 3 times more apples than Allan. Allan has 9 apples. How many apples does Marin have?

   ....

2. Michele's garden has 5 rows of pumpkins. Each row has 4 pumpkins. How many pumpkins does Michele have in all?

   ....

3. Paul swims 2 laps every day. How many laps will Paul swim in 10 days?

   ....

4. If there are 4 peaches in each box and there is 1 boxes, how many peaches are there in total?

   ....

5. Donald can cycle 3 miles per hour. How far can Donald cycle in 3 hours?

   ....

6. Marin's garden has 6 rows of pumpkins. Each row has 10 pumpkins. How many pumpkins does Marin have in all?

   ....

7. Marcie has 3 times more pears than Ellen. Ellen has 10 pears. How many pears does Marcie have?

   ....

1. Steven can cycle 6 miles per hour. How far can Steven cycle in 6 hours?

2. If there are 7 apples in each box and there are 8 boxes, how many apples are there in total?

3. Marin's garden has 1 rows of pumpkins. Each row has 5 pumpkins. How many pumpkins does Marin have in all?

4. Jennifer swims 4 laps every day. How many laps will Jennifer swim in 7 days?

5. Marcie has 3 times more balls than Michele. Michele has 6 balls. How many balls does Marcie have?

6. Ellen's garden has 2 rows of pumpkins. Each row has 4 pumpkins. How many pumpkins does Ellen have in all?

7. Marin swims 9 laps every day. How many laps will Marin swim in 7 days?

DIVISION

1. $3 \div 1 =$ ....

2. $16 \div 1 =$ ....

3. $32 \div 2 =$ ....

4. $1 \div 1 =$ ....

5. $7 \div 1 =$ ....

6. $36 \div 2 =$ ....

7. $20 \div 2 =$ ....

8. $18 \div 1 =$ ....

9. $18 \div 2 =$ ....

10. $19 \div 1 =$ ....

11. $13 \div 1 =$ ....

12. $11 \div 1 =$ ....

13. $28 \div 2 =$ ....

14. $26 \div 2 =$ ....

15. $4 \div 2 =$ ....

16. $6 \div 1 =$ ....

17. $2 \div 2 =$ ....

18. $14 \div 1 =$ ....

19. $30 \div 2 =$ ....

20. $17 \div 1 =$ ....

21. $6 \div 2 =$ ....

22. $34 \div 2 =$ ....

23. $4 \div 1 =$ ....

24. $10 \div 2 =$ ....

1. $8 \div 2 =$ ....

2. $21 \div 3 =$ ....

3. $57 \div 3 =$ ....

4. $22 \div 2 =$ ....

5. $14 \div 2 =$ ....

6. $27 \div 3 =$ ....

7. $24 \div 3 =$ ....

8. $12 \div 3 =$ ....

9. $18 \div 3 =$ ....

10. $15 \div 3 =$ ....

11. $54 \div 3 =$ ....

12. $20 \div 2 =$ ....

13. $48 \div 3 =$ ....

14. $51 \div 3 =$ ....

15. $28 \div 2 =$ ....

16. $60 \div 3 =$ ....

17. $30 \div 3 =$ ....

18. $12 \div 2 =$ ....

19. $36 \div 2 =$ ....

20. $42 \div 3 =$ ....

21. $45 \div 3 =$ ....

22. $2 \div 2 =$ ....

23. $30 \div 2 =$ ....

24. $36 \div 3 =$ ....

1. **40 ÷ 4 =**

2. **48 ÷ 4 =**

3. **9 ÷ 3 =**

4. **57 ÷ 3 =**

5. **36 ÷ 3 =**

6. **12 ÷ 3 =**

7. **18 ÷ 3 =**

8. **42 ÷ 3 =**

9. **60 ÷ 3 =**

10. **39 ÷ 3 =**

11. **76 ÷ 4 =**

12. **27 ÷ 3 =**

13. **32 ÷ 4 =**

14. **64 ÷ 4 =**

15. **4 ÷ 4 =**

16. **12 ÷ 4 =**

17. **54 ÷ 3 =**

18. **20 ÷ 4 =**

19. **30 ÷ 3 =**

20. **80 ÷ 4 =**

21. **68 ÷ 4 =**

22. **48 ÷ 3 =**

23. **6 ÷ 3 =**

24. **8 ÷ 4 =**

1. $32 \div 4 =$ ....

2. $90 \div 5 =$ ....

3. $85 \div 5 =$ ....

4. $40 \div 4 =$ ....

5. $36 \div 4 =$ ....

6. $64 \div 4 =$ ....

7. $12 \div 4 =$ ....

8. $60 \div 4 =$ ....

9. $35 \div 5 =$ ....

10. $30 \div 5 =$ ....

11. $56 \div 4 =$ ....

12. $48 \div 4 =$ ....

13. $4 \div 4 =$ ....

14. $44 \div 4 =$ ....

15. $10 \div 5 =$ ....

16. $15 \div 5 =$ ....

17. $8 \div 4 =$ ....

18. $70 \div 5 =$ ....

19. $55 \div 5 =$ ....

20. $68 \div 4 =$ ....

21. $80 \div 5 =$ ....

22. $20 \div 5 =$ ....

23. $52 \div 4 =$ ....

24. $28 \div 4 =$ ....

1. $42 \div 6 =$ ....

2. $30 \div 5 =$ ....

3. $90 \div 5 =$ ....

4. $24 \div 6 =$ ....

5. $5 \div 5 =$ ....

6. $84 \div 6 =$ ....

7. $50 \div 5 =$ ....

8. $60 \div 6 =$ ....

9. $48 \div 6 =$ ....

10. $65 \div 5 =$ ....

11. $45 \div 5 =$ ....

12. $78 \div 6 =$ ....

13. $95 \div 5 =$ ....

14. $18 \div 6 =$ ....

15. $114 \div 6 =$ ....

16. $6 \div 6 =$ ....

17. $60 \div 5 =$ ....

18. $12 \div 6 =$ ....

19. $66 \div 6 =$ ....

20. $20 \div 5 =$ ....

21. $108 \div 6 =$ ....

22. $35 \div 5 =$ ....

23. $90 \div 6 =$ ....

24. $55 \div 5 =$ ....

1. $30 \div 6 =$ ....

2. $108 \div 6 =$ ....

3. $78 \div 6 =$ ....

4. $48 \div 6 =$ ....

5. $24 \div 6 =$ ....

6. $91 \div 7 =$ ....

7. $126 \div 7 =$ ....

8. $114 \div 6 =$ ....

9. $140 \div 7 =$ ....

10. $28 \div 7 =$ ....

11. $98 \div 7 =$ ....

12. $54 \div 6 =$ ....

13. $12 \div 6 =$ ....

14. $119 \div 7 =$ ....

15. $18 \div 6 =$ ....

16. $6 \div 6 =$ ....

17. $21 \div 7 =$ ....

18. $112 \div 7 =$ ....

19. $84 \div 6 =$ ....

20. $72 \div 6 =$ ....

21. $56 \div 7 =$ ....

22. $36 \div 6 =$ ....

23. $105 \div 7 =$ ....

24. $66 \div 6 =$ ....

1. $48 \div 8 =$

2. $40 \div 8 =$

3. $98 \div 7 =$

4. $28 \div 7 =$

5. $91 \div 7 =$

6. $119 \div 7 =$

7. $49 \div 7 =$

8. $14 \div 7 =$

9. $35 \div 7 =$

10. $128 \div 8 =$

11. $70 \div 7 =$

12. $120 \div 8 =$

13. $104 \div 8 =$

14. $136 \div 8 =$

15. $42 \div 7 =$

16. $105 \div 7 =$

17. $112 \div 8 =$

18. $56 \div 8 =$

19. $112 \div 7 =$

20. $80 \div 8 =$

21. $56 \div 7 =$

22. $21 \div 7 =$

23. $126 \div 7 =$

24. $84 \div 7 =$

1. **96 ÷ 8 =**

2. **45 ÷ 9 =**

3. **144 ÷ 9 =**

4. **90 ÷ 9 =**

5. **135 ÷ 9 =**

6. **48 ÷ 8 =**

7. **72 ÷ 9 =**

8. **162 ÷ 9 =**

9. **36 ÷ 9 =**

10. **144 ÷ 8 =**

11. **27 ÷ 9 =**

12. **72 ÷ 8 =**

13. **117 ÷ 9 =**

14. **120 ÷ 8 =**

15. **16 ÷ 8 =**

16. **152 ÷ 8 =**

17. **99 ÷ 9 =**

18. **112 ÷ 8 =**

19. **18 ÷ 9 =**

20. **24 ÷ 8 =**

21. **80 ÷ 8 =**

22. **108 ÷ 9 =**

23. **88 ÷ 8 =**

24. **81 ÷ 9 =**

1. Brian is reading a book with 40 pages. If Brian wants to read the same number of pages every day, how many pages would Brian have to read each day to finish in 4 days?

2. Janet ordered 7 pizzas. The bill for the pizzas came to $70. What was the cost of each pizza?

3. A box of oranges weighs 2 pounds. If one oranges weighs 1 pounds, how many oranges are there in the box?

4. Sharon made 4 cookies for a bake sale. She put the cookies in bags, with 2 cookies in each bag. How many bags did she have for the bake sale?

5. How many 7 cm pieces of rope can you cut from a rope that is 21 cm long?

6. You have 42 pears and want to share them equally with 7 people. How many pears would each person get?

7. You have 8 peaches and want to share them equally with 8 people. How many peaches would each person get?

1. Sharon made 24 cookies for a bake sale. She put the cookies in bags, with 3 cookies in each bag. How many bags did she have for the bake sale?

2. Adam is reading a book with 70 pages. If Adam wants to read the same number of pages every day, how many pages would Adam have to read each day to finish in 7 days?

3. A box of apples weighs 24 pounds. If one apples weighs 6 pounds, how many apples are there in the box?

4. Sandra ordered 7 pizzas. The bill for the pizzas came to $49. What was the cost of each pizza?

5. You have 45 peaches and want to share them equally with 5 people. How many peaches would each person get?

6. How many 5 cm pieces of rope can you cut from a rope that is 25 cm long?

7. A box of balls weighs 70 pounds. If one balls weighs 10 pounds, how many balls are there in the box?

1. You have 20 peaches and want to share them equally with 5 people. How many peaches would each person get?

   ....

2. Steven is reading a book with 54 pages. If Steven wants to read the same number of pages every day, how many pages would Steven have to read each day to finish in 9 days?

   ....

3. Marcie ordered 10 pizzas. The bill for the pizzas came to $50. What was the cost of each pizza?

   ....

4. How many 8 cm pieces of rope can you cut from a rope that is 32 cm long?

   ....

5. Jennifer made 54 cookies for a bake sale. She put the cookies in bags, with 6 cookies in each bag. How many bags did she have for the bake sale?

   ....

6. A box of marbles weighs 20 pounds. If one marbles weighs 4 pounds, how many marbles are there in the box?

   ....

7. A box of balls weighs 40 pounds. If one balls weighs 10 pounds, how many balls are there in the box?

   ....

1. A box of pears weighs 60 pounds. If one pears weighs 10 pounds, how many pears are there in the box?

2. How many 10 cm pieces of rope can you cut from a rope that is 30 cm long?

3. Jackie ordered 10 pizzas. The bill for the pizzas came to $30. What was the cost of each pizza?

4. Allan is reading a book with 48 pages. If Allan wants to read the same number of pages every day, how many pages would Allan have to read each day to finish in 6 days?

5. Amy made 9 cookies for a bake sale. She put the cookies in bags, with 3 cookies in each bag. How many bags did she have for the bake sale?

6. You have 90 plums and want to share them equally with 10 people. How many plums would each person get?

7. Jennifer made 36 cookies for a bake sale. She put the cookies in bags, with 6 cookies in each bag. How many bags did she have for the bake sale?

Made in United States
Orlando, FL
30 June 2023

34644291R00033